ANCIENT WORLDS

ANCIENT INNOVATIONS THAT SHAPED HISTORY

Martin Gitlin

45TH PARALLEL PRESS

Published in the United States of America by Cherry Lake Publishing Group
Ann Arbor, Michigan
www.cherrylakepublishing.com

Reading Adviser: Beth Walker Gambro, MS, Ed., Reading Consultant, Yorkville, IL

Photo Credits: © richcano/iStock, cover, title page; © duncan1890/iStock, 4; © widiawanwastu/Shutterstock, 7; © Ali A Suliman/Shutterstock, 8; © Fernando Cortes/Shutterstock, 11; © Marquisphoto/Shutterstock, 12; © Stephen Bridger/Shutterstock, 14; © Grafissimo/iStock, 16; Louis Joseph LeBrun, Public domain, via Wikimedia Commons, 18; © Ihor Bondarenko/Shutterstock, 20; © 300dpi/Shutterstock, 22; © Iren Key/Shutterstock, 24; © saiko3p/Shutterstock, 27; © View Stock/Alamy Stock Photo, 28; National Library of Portugal, Public domain, via Wikimedia Commons, 30

Graphic Element Credits: Graphic Element Credits: Cover, multiple interior pages: © Andrey_Kuzmin/Shutterstock, © cajoer/Shutterstock, © GUSAK OLENA/Shutterstock, © Eky Studio/Shutterstock

45TH Parallel Press is an imprint of Cherry Lake Publishing Group.

Library of Congress Cataloging-in-Publication Data has been filed and is available at catalog.loc.gov.

Cherry Lake Publishing Group would like to acknowledge the work of the Partnership for 21st Century Learning, a Network of Battelle for Kids. Please visit http://www.battelleforkids.org/networks/p21 for more information.

Printed in the United States of America

Note from publisher: Websites change regularly, and their future contents are outside of our control. Supervise children when conducting any recommended online searches for extended learning opportunities.

About the Author
Martin Gitlin is an educational book author based in Connecticut. He won more than 45 awards as a newspaper journalist from 1991 to 2002. Included was a first-place award from the Associated Press. That organization voted him as one of the top four feature writers in Ohio in 2002. Martin has had about 250 books published since 2006. Most of them were written for students. He has authored many books about history.

TABLE OF CONTENTS

INTRODUCTION

Ancient people needed food. They **scavenged** and hunted. Scavenged means searched for food to eat. They did this to stay alive. New inventions made getting food easier.

Rocket ships soar to the Moon. Trains reach high speeds. They can go up to 361 miles (581 kilometers) per hour. The internet gives us information. Doctors make sick hearts healthy. Modern times are amazing. We've benefited from many inventions.

Innovation never stops. This is when people think of new ideas. Inventions became part of normal living. We need new inventions. Flying cars could replace rolling ones. People might vacation on Mars. Maybe cancer will be cured. Robots might do our chores. **Drones** will deliver our store orders. Drones are robots that fly without pilots.

People have always been curious. They want to improve their lives. They want to make things easier. So they went to work. They made great inventions. They innovated.

Ancient people lived thousands of years ago. They had tough lives. They didn't live long. They died from sicknesses easily treated today. They suffered from severe weather. They had to find food. Roads were poor. Most people walked everywhere they went. It took longer to travel far.

Life got easier. It got more enjoyable. That is because of innovation. Hard jobs were made simpler. Inventions improve communication. They improve travel. They improve health. They create better food. They give people entertainment.

Inventions have moved the world forward. The wheel changed travel. **Structures** were built using new materials. Structures are things that are built. Letters were written with ink and paper. Foods like chocolate were added to the menu. The compass helped people sail to new lands.

Innovation changed the ancient world. And it keeps changing today's world.

Ancient people told stories. Storytelling was the main form of entertainment. A lot of storytelling happened in ancient theaters.

CHAPTER ONE

Ancient people only had a few ways to travel. People used donkeys, camels, and horses.

Rolling Around: The First Wheel

People still walk. They walk to the store. They walk in the park. They walk around the block. Most people choose to walk. They want exercise. They want fresh air. But imagine if walking was the only option. Ancient people didn't have bikes. They didn't have cars. They didn't have planes.

That was because there were no wheels. People could walk. Or they could ride animals. That was better for long trips. But people couldn't move fast. Travel was limited. It took longer to reach faraway places.

People sometimes say, "**Necessity** is the mother of invention." Necessity means needing something. Ancient people needed easier ways to travel. Yet other inventions came first. Even the flute was invented before the wheel!

Things changed between 5000 and 3500 BCE. That's when Sumerians invented the wheel. Sumerians lived in Mesopotamia. Today, that area is Iraq. The first wheels were solid wooden **discs**. Discs are flat, thin, round objects. The wood was cut from tree trunks.

These first wheels weren't for travel. They were used by **potters**. Potters made pots, plates, and more. Wheels were used to shape clay. People started using wheels for travel around 3400 BCE.

Two new ideas emerged. One was cutting smaller pieces of wood. This made wheels lighter and faster. The other was the invention of the **axle**. Axles are rods. The rod fits in a hole inside the wheel. Turning the axle turned the wheel. Axles were used in carts and **chariots**. Chariots are carts. They have two wheels. They're pulled by animals.

Chariot races took place in ancient Rome. People watched them for entertainment.

CHAPTER TWO

Drivers on modern highways can thank ancient people. Ancient people invented concrete.

Building Better with Concrete

Modern people take much for granted. They drive on roads. They cross bridges. They park in parking lots. They live in homes. They work and shop in buildings. And they think nothing of it.

All of these things require **concrete**. Concrete is a strong building material. Some buildings are made of steel or glass. But they still have concrete bases. Concrete is everywhere. But ancient concrete was made differently from modern concrete.

Concrete was first used around 6500 BCE. It was first used in ancient Syria and Jordan. It was found in floors. It was found in homes.

Concrete's wide use began 2,100 years ago. Ancient Romans made concrete popular. They did this around 150 BCE. The concrete they made was tougher.

The ancient Romans mixed limestone with volcanic ash. They added water. That formed **mortar**. Mortar is like a paste. The ancient Romans packed the mortar with brick or volcanic rock. That mixture was strong. It formed sturdy roads. It formed sturdy bridges. It formed sturdy buildings.

The Colosseum is made of concrete. That is where **gladiators** battled. Gladiators are trained fighters. The Pantheon temple is also made of concrete. Both are still standing. The ancient Romans knew how to make things that last.

It took close to 10 years to build the Colosseum. This was very fast for ancient times.

SHARING THE NEWS

An early form of paper was invented in ancient Egypt. But the first newspaper came from ancient Rome. And it was not printed on paper. It was carved on metal or stone.

It was called the *Acta Diurna*. That means "daily acts." It first appeared before 59 CE. Readers learned about political events. They learned about social events. They learned about military victories. They learned where gladiators were battling. They learned about local births. They learned about deaths.

The news was posted where people could see it. It was posted in busy areas. An example is the Roman Forum. The Roman Forum was at the center of the city. It's where public meetings took place. Publishing news of the Roman Senate began later. The *Acta Diurna* continued during the Roman Empire. The empire began about 27 BCE. It ended in 476 CE.

CHAPTER THREE

The Minoans were known for their amazing artwork. This was during the Bronze Age. The Bronze Age was from 3300 BCE to 1200 BCE.

Giving Birth to Math

Math can be as easy as 1 plus 1. The first humans understood that. Real innovation began 4,000 years ago. That is when math became useful.

Math can be traced back to Mesopotamia. People there solved math problems. They used math to divide land. They divided land for farming.

The Minoans also used math. They lived on the island of Crete. Crete is part of modern Greece. Early Minoan wall paintings date to around 2000 BCE. They show **geometric** figures. Geometry includes shapes.

Greek experts advanced geometry. Among them were Thales and Pythagoras. They introduced geometry. They taught it to others. Euclid helped put it into practice.

Plato and Aristotle were early **philosophers**. Philosophers are people who seek wisdom They used geometry to search for truth. They studied logic. Geometry also inspired Archimedes. Archimedes is the "father of math." He came up with many math concepts.

Sumerians worked with numbers. They did this around 2500 BCE. They based time on the number 60. (There are 60 seconds in a minute. There are 60 minutes in an hour.) Math advanced in ancient Egypt. The Egyptians worked with basic numbers. They added. They subtracted. They divided. They multiplied.

So, who invented math? That's hard to answer. Many ancient peoples played a part. Math is needed for many inventions.

Socrates was the first great Greek philosopher. He taught Plato. Plato taught Aristotle.

MYSTERY SOLVED OR UNSOLVED?

Who drank the first glass of beer? It's thought to have been made on the Arabian Peninsula. That thinking changed in 2013. Ancient pottery was found in China that year. The pottery was in the Henan province. This province is in north-central China. The pottery had traces of alcohol. It was 9,000 years old. That was 1,000 years before the Arabians brewed beer. But was it really beer? No one knows for sure.

Alcohol is known as "jiu" in China. Studies showed that ancient Chinese drank beer. Mentions of alcohol were even inscribed on animal bones. These were written during the Shang Dynasty. That was between 1600 and 1046 BCE.

Other ancient cultures had alcohol. Ancient Greeks made mead. Mead was made from honey and water. Indigenous Americans made drinks from corn and grapes.

CHAPTER FOUR

Papyrus is known for its crisscross pattern. Egyptians did this by laying strips of papyrus side by side. The first layer was vertical. The second layer was horizontal.

Inventing Paper and Pens

It was ancient times. There were no typewriters. There were no laptops. How did people share information? How did they pass messages? They used pens. They wrote words on paper.

Modern pens and paper started in ancient Egypt. It was around 3000 B.C. Ancient Egyptians noticed the papyrus plant. Papyrus grew along the Nile River. Its stems could be torn off. People wove the strips together. They weighted them down. The pressure turned the woven strips into a strong, thin sheet. This made paper.

People still needed a writing tool. Ancient Egyptians also took care of that.

Ancient Egyptians cut reeds. The reeds were strong. They filled the reeds with ink. Ink was made with soot. Soot was mixed with beeswax. It was mixed with vegetable gum.

Ancient Egyptians made **hieroglyphics**. Hieroglyphics is a system of picture writing. Many are still intact. They can still be read 5,000 years later.

Ancient Egypt had tombs and temples. Hieroglyphics were found on the walls.

ANCIENT RULES

The ancient Mayans of Central America had strict laws. Each state was ruled by a king. People believed kings worked for ancient Mayan gods. They believed gods gave kings the right to rule. But kings didn't rule alone. Powerful councils ran the government. They were elected by noblemen. These men were from the wealthy classes.

Most crimes were punished harshly. The ancient Mayans didn't have prisons. Some criminals were killed. Others were placed into slavery. Those who committed lesser crimes were fined. They had to pay fees. The guilty sometimes had their heads shaved. That was a sign of shame in ancient Mayan culture.

People accused of crimes appeared in court. Kings sometimes served as judges. Witnesses testified. They spoke for or against the accused. The guilty were punished immediately.

CHAPTER FIVE

Mayans developed the science of **astronomy**. This is the study of the universe beyond Earth. They developed calendar systems.

Bringing Chocolate to the World

What is the tastiest treat? Many say it's chocolate. Chocolate is bitter alone. But it is sweet with sugar. It makes people happy. Mesoamericans introduced chocolate to the world.

Mesoamericans lived in what is now Mexico and Central America. They formed several advanced **civilizations**. Civilizations are human societies. They have governments. They have cultures.

Ancient Mayans and Aztecs were Mesoamericans. Mayans lived in southern Mexico, Guatemala, and northern Belize. Aztecs lived in northern Mexico.

Mesoamericans made a great discovery. They did this about 3,000 years ago. They found a bean on the cocoa tree.

Cocoa trees were sacred. The people thought gods gave them these trees.

The cocoa bean was tasty. But it was also useful. Mesoamericans learned that. They realized it gave them energy. It also improved their moods.

Ancient Mayans loved cocoa beans. These beans became more than a yummy treat. They could be used as **currency**. Currency is a money system. Cocoa beans were traded for valuable things.

Spanish explorers arrived in the 1500s. They were seeking gold and silver. But they also found cocoa beans. They knew a good thing when they tasted it. They loved chocolate. They returned to Europe. They brought cocoa beans. That started a chocolate craze. It began nearly 500 years ago. And it has never ended. Chocolate is loved all over the world.

Fats in chocolate give energy to the body. Caffeine in coffee gives energy to the brain.

CHAPTER SIX

Some ancient Chinese compasses, like this one, were made with lodestone. It is naturally magnetic.

Creating the Compass

Today, most people use GPS. GPS is Global Positioning System. It tells us direction. Long ago, people used stars. The stars helped them **navigate**. Navigate means to direct travel. But what if it was cloudy? The invention of the compass helped. A compass could tell travelers where they were going. It could give them direction. But that was not its first use.

Nobody is sure when the compass was invented. It may have been between 200 BCE and 100 CE. But we know where it was invented. It was invented in Ancient China. It was first used in **Feng Shui**. That is a spiritual practice. It's about good energy. It uses the layout of buildings. It uses the layout of objects.

Ancient Chinese traders wanted to explore other lands. They began sailing ships around 1000 CE. The compass helped them navigate.

The Arab world is mainly in western Asia and northern Africa. Arab traders sailed to China. They heard about the compass. They brought it to Europe. That inspired Europe to explore.

Using the compass took time. The compass was first used as a backup. Explorers looked at other landmarks as guides. They used the Sun. They used the stars. They used the compass when they couldn't see landmarks.

The compass became more reliable. People learned how to read it better. Explorers used it to make discoveries.

Ferdinand Magellan was an explorer. He was from Portugal. He used the compass. He started traveling in 1519.

VERY IMPORTANT PEOPLE

Little is written about ancient Korea. But ancient Koreans invented something very important. Modern life depends on it. Ancient Koreans invented central heating. They started heating their homes around 5000 BCE. Their method was simple. But it kept them warm. It was called ondol. The Koreans guided heat from fires used for cooking. The heat traveled through tunnels. These tunnels were under the floor.

Ancient Romans improved this method. That was about 5,000 years later. They built hollow chambers. These chambers were between the ground and living space. Hot air rose from the fires. It flowed through the chambers. It heated the spaces above.

Europeans later returned to basic methods. They used indoor fireplaces to heat their homes. Modern central heating dates to 1919. It was invented by Alice Parker. Parker was a Black inventor.

GLOSSARY

astronomy (uh-STRAH-nuh-mee) the study of the universe beyond Earth

axle (AK-suhl) a rod or spindle passing through the center of a wheel on which it turns

chariots (CHAIR-ee-uhts) 2-wheeled horse-drawn vehicles used in ancient warfare and racing

civilizations (sih-vuh-luh-ZAY-shuhnz) human societies at specific times and places that have government and cultures

concrete (KAHN-kreet) a hard strong building material made by mixing a cementing material with sand, gravel, and water

currency (KUHR-uhn-see) a system of money that is used to buy and sell things

discs (DISKS) flat, thin, round objects

drones (DROHNZ) any small uncrewed airborne devices that are guided remotely

Feng Shui (FUHNG SHWEE) the building layout and placement of objects to achieve a desired spiritual result

geometric (jee-uh-MEH-trik) made of lines and shapes

gladiators (GLAh-dee-ay-tuhrz) ancient Romans who fought animals and other men

graphite (GRA-fiyt) a soft, shiny black carbon that is used in making lead pencils

hieroglyphics (hye-ruh-GLIH-fiks) ancient writing that used pictures and symbols

innovation (ih-nuh-VAY-shuhn) when people think of creative new ideas

mortar (MOHR-tuhr) material made from lime, sand, and water

navigate (NAH-vuh-gayt) to travel or steer to or around something

necessity (nih-SEH-suh-tee) something that is essential

philosophers (fuh-LAH-suh-fuhrz) people who seek wisdom

potters (PAH-tuhrz) craftspeople who make pottery such as pots, bowls, plates, etc.

scavenged (SKAA-vuhnjd) searched for food to eat

structures (STRUHK-churz) things built from a number of parts joined together in a certain way

LEARN MORE

Loh-Hagan, Virginia. *Alice Parker and the Furnace*. Cherry Lake Publishing, 2018.

O'Neill, Bill. *Amazing Inventions that Changed the World: The True Stories about the Revolutionary and Accidental Inventions that Changed Our World*. Sheridan, WY: LAK Publishing, 2022.

Professor Beaver. *Ancient China's Inventions, Technology and Engineering—Ancient History Book for Kids*. Professor Beaver, 2017.

Search online with an adult:

Ancient Innovations for Kids. Mr. Donn's Ancient History.

INDEX